Product Design Logbook

PRODUCT DESIGN

LOGBOOK

An Inventor's Notebook

Volume Number _____

Property of:

_____
Name

_____
Address

_____

If found, please contact

_____

ISBN 10 Digit: 1-933598-92-1

ISBN 13 Digit: 978-1-933598-92-5

Published by Johnson & Hunter, Inc.
www.johnsonhunter.com

Cover Graphic: Art'nLera/Shutterstock.com

Contents

Project   Page

Contents

Project   Page

Contents

Project   Page

Contents

Project   Page

Contents

Project   Page

_____

_____

_____

_____

_____

_____

_____

_____

_____

_____

_____

_____

_____

_____

_____

_____

_____

_____

_____

Contents

Project   Page

Contents

Project   Page

Contents

Project   Page

Preface

Ideas can pop up when you least expect it. The purpose of this logbook is to provide a handy journal with all the elements you need to properly record your ideas, inventions and innovations. The grid provides 1/4 inch spacing and the Contents section enables you to record a quick reference to locate your projects and specific notes or versions of the projects.

At the back of the logbook, you will find space to write names and addresses of important contacts who may be relevant to the projects you have logged in this journal. Additionally, I have provided some recommended reading you may find helpful as you embark upon and continue your journey.

Here are some best practices to keep in mind as you log your ideas and concepts:

- Use permanent ink, not pencil.

- Do not skip pages.

- Date each entry.

- Have someone you know who is unrelated to you or your company witness your entry, preferably the same day that you created the concept.

- Use the same ink pen for a single page to avoid a perception that an entry was altered.

- You may want to keep separate logbooks for each idea on the outside chance that at some point you may have to defend your position as first to invent in litigation. If you have to produce documentation, then you will have to redact your other concepts to avoid providing sensitive information into public record. We also have other editions that can be used as a logbook for a single concept. For more information, visit **www.productdesignlogbook.com** for other editions.

- As you create, make notes about how your product fills a need, enhances an existing product, or improves production.

- Write down your thoughts about how your product will help a targeted buyer -- whether it is the customer who buys the product or a company who licenses your idea -- increase their income, reduce their expenses, manage their risks, improve their productivity, or feel better by buying your product. These will be key points to help you sell or license your idea and help you prepare for presentations.

- When you make notes about your research, write as much information as possible about your source (i.e., web addresses; article title, date, and publisher; magazine editions; press release information; database sources; third-party research title, date, and location, etc.).

These best practices are only a handful, but they represent important ones.

May you achieve much success as you invent, design, research, and sell your ideas!

Renee DiModica

Project Title

Date

Witnessed by: _____

Print Name: _____

Telephone No: _____

Project Title	Date

Witnessed by: _____

Print Name: _____

Telephone No: _____

Project Title	Date

Witnessed by: _____

Print Name: _____

Telephone No: _____

Project Title

Date

Witnessed by: _____

Print Name: _____

Telephone No: _____

Project Title

Date

Witnessed by: _____

Print Name: _____

Telephone No: _____

Project Title

Date

Witnessed by: _____

Print Name: _____

Telephone No: _____

Project Title

Date

Witnessed by: _____

Print Name: _____

Telephone No: _____

Project Title

Date

Witnessed by: _____

Print Name: _____

Telephone No: _____

Project Title

Date

Witnessed by: _____

Print Name: _____

Telephone No: _____

Project Title

Date

Witnessed by: _____

Print Name: _____

Telephone No: _____

Project Title

Date

Witnessed by: _____

Print Name: _____

Telephone No: _____

Project Title

Date

Witnessed by: _____

Print Name: _____

Telephone No: _____

Project Title

Date

Witnessed by: _____

Print Name: _____

Telephone No: _____

Project Title

Date

Witnessed by: _____

Print Name: _____

Telephone No: _____

Project Title

Date

Witnessed by: _____

Print Name: _____

Telephone No: _____

Project Title

Date

Witnessed by: _____

Print Name: _____

Telephone No: _____

Project Title

Date

Witnessed by: _____

Print Name: _____

Telephone No: _____

Project Title

Date

Witnessed by: _____

Print Name: _____

Telephone No: _____

Project Title

Date

Witnessed by: _____

Print Name: _____

Telephone No: _____

Project Title

Date

Witnessed by: _____

Print Name: _____

Telephone No: _____

Project Title	Date

Witnessed by: _____

Print Name: _____

Telephone No: _____

Project Title	Date

Witnessed by: _____

Print Name: _____

Telephone No: _____

Project Title

Date

Witnessed by: _____

Print Name: _____

Telephone No: _____

Project Title

Date

Witnessed by: _____

Print Name: _____

Telephone No: _____

Project Title	Date

Witnessed by: _____

Print Name: _____

Telephone No: _____

Project Title

Date

Witnessed by: _____

Print Name: _____

Telephone No: _____

Project Title

Date

Witnessed by: _____

Print Name: _____

Telephone No: _____

Project Title

Date

Witnessed by: _____

Print Name: _____

Telephone No: _____

Project Title

Date

Witnessed by: _____

Print Name: _____

Telephone No: _____

Project Title

Date

Witnessed by: _____

Print Name: _____

Telephone No: _____

Project Title

Date

Witnessed by: _____

Print Name: _____

Telephone No: _____

Project Title

Date

Witnessed by: _____

Print Name: _____

Telephone No: _____

Project Title

Date

Witnessed by: _____

Print Name: _____

Telephone No: _____

Project Title	Date

Witnessed by: _____

Print Name: _____

Telephone No: _____

Project Title	Date

Witnessed by: _____

Print Name: _____

Telephone No: _____

Project Title

Date

Witnessed by: _____

Print Name: _____

Telephone No: _____

Project Title

Date

Witnessed by: _____

Print Name: _____

Telephone No: _____

Project Title

Date

Witnessed by: _____

Print Name: _____

Telephone No: _____

Project Title

Date

Witnessed by: _____

Print Name: _____

Telephone No: _____

Project Title

Date

Witnessed by: _____

Print Name: _____

Telephone No: _____

Project Title

Date

Witnessed by: _____

Print Name: _____

Telephone No: _____

Project Title

Date

Witnessed by: _____

Print Name: _____

Telephone No: _____

Project Title

Date

Witnessed by: _____

Print Name: _____

Telephone No: _____

Project Title

Date

Witnessed by: _____

Print Name: _____

Telephone No: _____

Project Title	Date

Witnessed by: _____

Print Name: _____

Telephone No: _____

Project Title

Date

Witnessed by: _____

Print Name: _____

Telephone No: _____

Project Title

Date

Witnessed by: _____

Print Name: _____

Telephone No: _____

Project Title

Date

Witnessed by: _____

Print Name: _____

Telephone No: _____

Project Title

Date

Witnessed by: _____

Print Name: _____

Telephone No: _____

Project Title

Date

Witnessed by: _____

Print Name: _____

Telephone No: _____

Project Title

Date

Witnessed by: _____

Print Name: _____

Telephone No: _____

Project Title

Date

Witnessed by: _____

Print Name: _____

Telephone No: _____

Project Title	Date

Witnessed by: _____

Print Name: _____

Telephone No: _____

Project Title

Date

Witnessed by: _____

Print Name: _____

Telephone No: _____

Project Title	Date

Witnessed by: _____

Print Name: _____

Telephone No: _____

Project Title

Date

Witnessed by: _____

Print Name: _____

Telephone No: _____

Project Title	Date

Witnessed by: _____

Print Name: _____

Telephone No: _____

Project Title

Date

Witnessed by: _____

Print Name: _____

Telephone No: _____

Project Title

Date

Witnessed by: _____

Print Name: _____

Telephone No: _____

Project Title

Date

Witnessed by: _____

Print Name: _____

Telephone No: _____

Project Title	Date

Witnessed by: _____

Print Name: _____

Telephone No: _____

Project Title

Date

Witnessed by: _____

Print Name: _____

Telephone No: _____

Project Title

Date

Witnessed by: _____

Print Name: _____

Telephone No: _____

Project Title

Date

Witnessed by: _____

Print Name: _____

Telephone No: _____

Project Title	Date

Witnessed by: _____

Print Name: _____

Telephone No: _____

Project Title

Date

Witnessed by: _____

Print Name: _____

Telephone No: _____

Project Title

Date

Witnessed by: _____

Print Name: _____

Telephone No: _____

Project Title

Date

Witnessed by: _____

Print Name: _____

Telephone No: _____

Project Title

Date

Witnessed by: _____

Print Name: _____

Telephone No: _____

Project Title

Date

Witnessed by: _____

Print Name: _____

Telephone No: _____

Project Title

Date

Witnessed by: _____

Print Name: _____

Telephone No: _____

Project Title	Date

Witnessed by: _____

Print Name: _____

Telephone No: _____

Project Title	Date

Witnessed by: _____

Print Name: _____

Telephone No: _____

Project Title

Date

Witnessed by: _____

Print Name: _____

Telephone No: _____

Project Title	Date

Witnessed by: _____

Print Name: _____

Telephone No: _____

Project Title

Date

Witnessed by: _____

Print Name: _____

Telephone No: _____

Project Title

Date

Witnessed by: _____

Print Name: _____

Telephone No: _____

Project Title	Date

Witnessed by: _____

Print Name: _____

Telephone No: _____

Project Title

Date

Witnessed by: _____

Print Name: _____

Telephone No: _____

Project Title	Date

Witnessed by: _____

Print Name: _____

Telephone No: _____

Project Title

Date

Witnessed by: _____

Print Name: _____

Telephone No: _____

Project Title

Date

Witnessed by: _____

Print Name: _____

Telephone No: _____

Project Title	Date

Witnessed by: _____

Print Name: _____

Telephone No: _____

Project Title

Date

Witnessed by: _____

Print Name: _____

Telephone No: _____

Project Title	Date

Witnessed by: _____

Print Name: _____

Telephone No: _____

Project Title

Date

Witnessed by:

Print Name:

Telephone No:

Project Title	Date

Witnessed by: _____

Print Name: _____

Telephone No: _____

Project Title	Date

Witnessed by: _____

Print Name: _____

Telephone No: _____

Project Title

Date

Witnessed by: _____

Print Name: _____

Telephone No: _____

Project Title

Date

Witnessed by: _____

Print Name: _____

Telephone No: _____

Project Title

Date

Witnessed by: _____

Print Name: _____

Telephone No: _____

Project Title

Date

Witnessed by: _____

Print Name: _____

Telephone No: _____

Project Title

Date

Witnessed by: _____

Print Name: _____

Telephone No: _____

Project Title

Date

Witnessed by: _____

Print Name: _____

Telephone No: _____

Project Title	Date

Witnessed by: _____

Print Name: _____

Telephone No: _____

Project Title

Date

Witnessed by: _____

Print Name: _____

Telephone No: _____

Project Title	Date

Witnessed by: _____

Print Name: _____

Telephone No: _____

Project Title

Date

Witnessed by:

Print Name:

Telephone No:

Project Title

Date

Witnessed by: _____

Print Name: _____

Telephone No: _____

Project Title

Date

Witnessed by: _____

Print Name: _____

Telephone No: _____

Project Title

Date

Witnessed by: _____

Print Name: _____

Telephone No: _____

Project Title

Date

Witnessed by: _____

Print Name: _____

Telephone No: _____

Project Title	Date

Witnessed by: _____

Print Name: _____

Telephone No: _____

Project Title	Date

Witnessed by: _____

Print Name: _____

Telephone No: _____

Project Title

Date

Witnessed by: _____

Print Name: _____

Telephone No: _____

Project Title

Date

Witnessed by:

Print Name:

Telephone No:

Project Title	Date

Witnessed by: _____

Print Name: _____

Telephone No: _____

Project Title

Date

Witnessed by: _____

Print Name: _____

Telephone No: _____

Project Title

Date

Witnessed by: _____

Print Name: _____

Telephone No: _____

Project Title

Date

Witnessed by: _____

Print Name: _____

Telephone No: _____

Project Title

Date

Witnessed by: _____

Print Name: _____

Telephone No: _____

Project Title

Date

Witnessed by: _____

Print Name: _____

Telephone No: _____

Project Title

Date

Witnessed by: _____

Print Name: _____

Telephone No: _____

Project Title	Date

Witnessed by: _____

Print Name: _____

Telephone No: _____

Project Title	Date

Witnessed by: _____

Print Name: _____

Telephone No: _____

Project Title

Date

Witnessed by: _____

Print Name: _____

Telephone No: _____

Project Title

Date

Witnessed by:

Print Name:

Telephone No:

Project Title	Date

Witnessed by: _____

Print Name: _____

Telephone No: _____

Project Title

Date

Witnessed by: _____

Print Name: _____

Telephone No: _____

Project Title	Date

Witnessed by: _____

Print Name: _____

Telephone No: _____

Project Title	Date

Witnessed by: _____

Print Name: _____

Telephone No: _____

Project Title	Date

Witnessed by: _____

Print Name: _____

Telephone No: _____

Project Title

Date

Witnessed by:

Print Name:

Telephone No:

Project Title	Date

Witnessed by: _____

Print Name: _____

Telephone No: _____

Project Title

Date

Witnessed by: _____

Print Name: _____

Telephone No: _____

Project Title

Date

Witnessed by:

Print Name:

Telephone No:

Project Title

Date

Witnessed by: _____

Print Name: _____

Telephone No: _____

Project Title	Date

Witnessed by: _____

Print Name: _____

Telephone No: _____

Project Title	Date

Witnessed by: _____

Print Name: _____

Telephone No: _____

Project Title	Date

Witnessed by: _____

Print Name: _____

Telephone No: _____

Project Title	Date

Witnessed by: _____

Print Name: _____

Telephone No: _____

Project Title	Date

Witnessed by: _____

Print Name: _____

Telephone No: _____

Project Title	Date

Witnessed by: _____

Print Name: _____

Telephone No: _____

Project Title	Date

Witnessed by: _____

Print Name: _____

Telephone No: _____

Project Title	Date

Witnessed by: _____

Print Name: _____

Telephone No: _____

Project Title

Date

Witnessed by: _____

Print Name: _____

Telephone No: _____

Project Title

Date

Witnessed by: _____

Print Name: _____

Telephone No: _____

Project Title

Date

Witnessed by: _____

Print Name: _____

Telephone No: _____

Project Title	Date

Witnessed by: _____

Print Name: _____

Telephone No: _____

Project Title

Date

Witnessed by: _____

Print Name: _____

Telephone No: _____

Project Title

Date

Witnessed by: _____

Print Name: _____

Telephone No: _____

Project Title	Date

Witnessed by: _____

Print Name: _____

Telephone No: _____

Project Title

Date

Witnessed by: ....

Print Name: ....

Telephone No: ....

Project Title

Date

Witnessed by: _____

Print Name: _____

Telephone No: _____

Project Title

Date

Witnessed by: _____

Print Name: _____

Telephone No: _____

Project Title

Date

Witnessed by: _____

Print Name: _____

Telephone No: _____

Project Title

Date

Witnessed by: _____

Print Name: _____

Telephone No: _____

Project Title

Date

Witnessed by: _____

Print Name: _____

Telephone No: _____

Project Title

Date

Witnessed by: _____

Print Name: _____

Telephone No: _____

Project Title

Date

Witnessed by: _____

Print Name: _____

Telephone No: _____

Project Title

Date

Witnessed by: _____

Print Name: _____

Telephone No: _____

Project Title

Date

Witnessed by: _____

Print Name: _____

Telephone No: _____

Project Title

Date

Witnessed by: _____

Print Name: _____

Telephone No: _____

Project Title

Date

Witnessed by: _____

Print Name: _____

Telephone No: _____

Project Title

Date

Witnessed by: _____

Print Name: _____

Telephone No: _____

Project Title	Date

Witnessed by: _____

Print Name: _____

Telephone No: _____

Project Title

Date

Witnessed by: _____

Print Name: _____

Telephone No: _____

Project Title

Date

Witnessed by: _____

Print Name: _____

Telephone No: _____

Project Title

Date

Witnessed by: _____

Print Name: _____

Telephone No: _____

Project Title	Date

Witnessed by: _____

Print Name: _____

Telephone No: _____

Contacts & Addresses

Name

Address

Phone   Mobile Phone

Notes

Name

Address

Phone   Mobile Phone

Notes

Name

Address

Phone   Mobile Phone

Notes

Name

Address

Phone   Mobile Phone

Notes

Name

Address

Phone   Mobile Phone

Notes

Name

Address

Phone   Mobile Phone

Notes

Name

Address

Phone   Mobile Phone

Notes

Name

Address

Phone   Mobile Phone

Notes

Name

Address

Phone   Mobile Phone

Notes

Name

Address

Home Phone   Mobile Phone

Notes

Recommended Reading

Inventing, Licensing, Selling:

Greiner, Lori. *Invent It, Sell It, Bank It!* Ballantine Books, 2014.

Key, Stephen with Colleen Sell. *One Simple Idea: Turn Your Dreams into a Licensing Goldmine, While Letting Others Do The Work*. McGraw Hill, 2011.

Drawing and Sketching:

Baskinger, Mark and William Bardel. *Drawing Ideas: A Hand-Drawn Approach for Better Design.* Watson-Guptill, 2013.

Robertson, Scott. *How to Draw: drawing and sketching objects and environments from your imagination.* Design Studio Press, 2013.

Graphic Design and Preparing a Sell Sheet:

Craig, James, William Bevington, and Irene Korol Scala. *Designing with Type, 5th Edition: The Essential Guide to Typography.* Watson-Guptill, 2006.

Eisemann, Leatrice. *Pantone Guide to Communicating With Color.* HOW Books, 2000.

Williams, Robin. *The Non-Designer's Design Book.* Peachpit Press, 2008.

Other reading:

Goldstein, Noah J., Steve J. Martin, and Robert B. Cialdini. *Yes!: 50 Scientifically Proven Ways to Be Persuasive.* Free Press, 2009.

Pink, Daniel H. *Drive: The Surprising Truth About What Motivates Us.* Riverhead Books, 2011.

Zadra, Dan. *5: Where Will You Be Five Years from Today?* Compendium, Inc., 2009.

Other Editions Available

at

www.productdesignlogbook.com

CPSIA information can be obtained
at www.ICGtesting.com
Printed in the USA
BVHW030804160720
583083BV00035B/68/J